# Hats On!

by Susan Markowitz Meredith

Look at all the hats.

One girl puts on a red hat.
Now she's a firefighter.

How many firefighter hats do you see?

Two boys put on brown hats.
They are cowboys.

How many cowboy hats are there?

Three girls put on white hats.
They are chefs now.
They are making cakes.

How many chef hats
do you see?

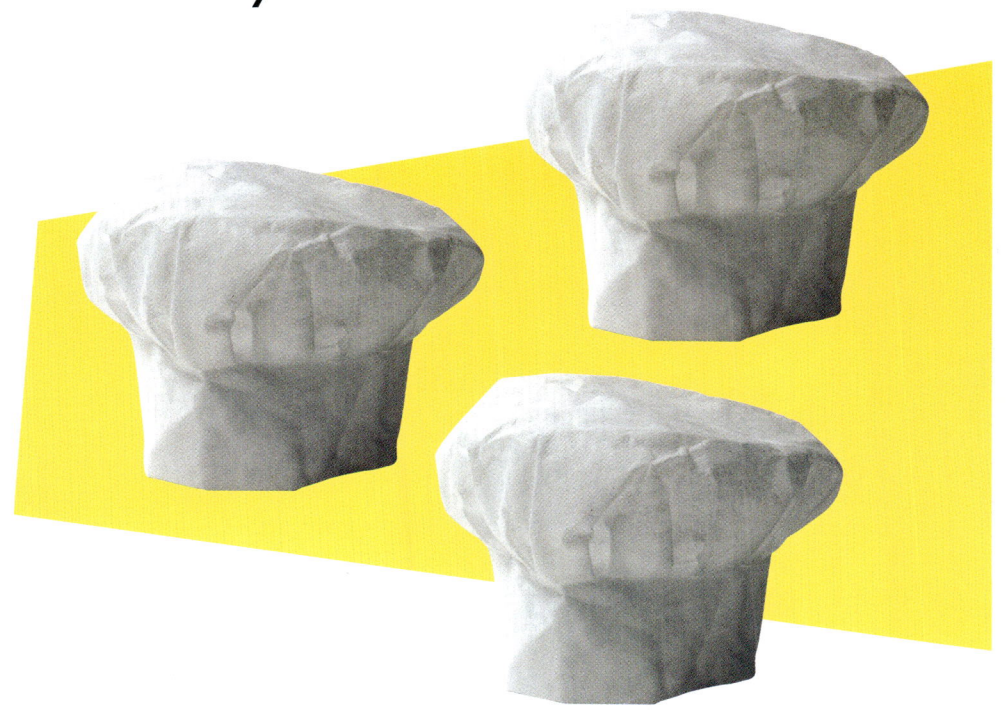

Four boys put on black hats. Now they are pirates. Aargh!

How many pirate hats are there?

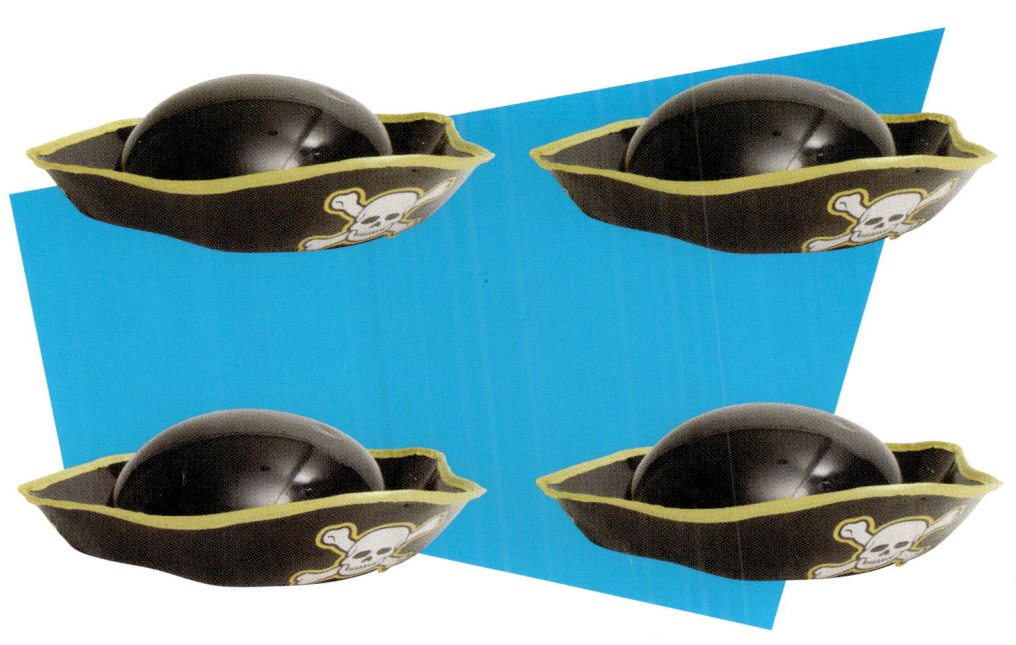

Five girls put on yellow hats.
They are builders now.

How many builder hats
do you see?

Six boys and girls put on clown hats. They are clowns at the circus.

How many clown hats are there?

It is fun to put on hats!